What is
FIBRE CHANNEL
?

third edition

ANCOT CORPORATION
115 Constitution Drive
Menlo Park, California 94025
Phone: (415) 322-5322
Fax: (415) 322-0455
Web site: http://www.ancot.com

COMPANY PROFILE

ANCOT Corporation designs and manufactures FIBRE CHANNEL and SCSI test equipment and other computer products, based on the Fibre Channel and SCSI interfaces for OEM markets.   ANCOT's test instruments and OEM products are used worldwide by leading computer companies. Applications include development, manufacturing, and repair. ANCOT's policy is to provide quality products, and to support its customers with qualified engineering support to maximize the return on their investment.

The book *WHAT IS FIBRE CHANNEL?* was written by Jan Dedek and Gary Stephens.

ISBN 0-9637439-1-0

CONTENTS

LIST OF FIGURES

Figure   **Page**

INTRODUCTION

Fibre Channel is the general name of an integrated set of standards being developed by the American National Standards Institute (ANSI) which defines new protocols for flexible information transfer. Fibre Channel development began in 1988 as an extension of work on the Intelligent Peripheral Interface (IPI) Enhanced Physical standard, and branched out in several directions.

The primary objectives are to provide:

- multiple physical interface types,
- a means to interconnect those various interface types,
- high-speed transfer of large amounts of information,
- separation of the logical protocol being transported from the physical interface,
- ability to transport multiple protocols over a common physical interface (perhaps concurrently),
- relief from the growing set of physical interfaces from which manufacturers must choose, many of which have limited general use.

Fibre Channel will be found in systems of many sizes. It first started appearing in workstations and in large mainframe systems. You may never see it in a low cost PC, but you might find it wherever you find power users of desktop systems, and users of desktop systems connecting to networks.

At present, you can find proprietary fiber optic interfaces on some desktop systems and workstations. The use of high-speed Fibre Channel interfaces will grow rapidly as their speed and functions become known to these users. This will include small work groups that have high-speed interconnect requirements for data interchange.

There are two basic peripheral protocols for device communication: *channels* and *networks*.   Traditionally, the term *channel* indicates a peripheral I/O interface to a host computer that transports large amounts of data between the host and the peripheral device. The processing overhead is kept to a minimum by handling data transfer in hardware, with little or no software involvement once an I/O operation begins. On the other hand, the term *network* indicates an I/O interface that usually supports many small transactions with higher overheads, usually the result of software involvement in the flow of information. Networks usually support host-to-host communication.

Channels

Channels usually operate in a closed, structured, and predictable environment where all devices that can communicate with a host are known in advance, and any change requires host software or configuration table changes. This advance knowledge of the configuration is very important in the high performance levels achieved by most channels.

The host system retains all the knowledge of the devices attached to it. Sometimes this is called a *master-slave* environment. Peripheral devices such as tapes, disks, and printers attach directly to the host system. The host is the master and these peripheral devices are the slaves.

Channels are used for transferring data. By data, we mean files of information which can be many thousands of bytes long. An important requirement for transferring data is error-free delivery, with transfer delay being a secondary consideration.

Networks

Networks operate in an open, unstructured, and essentially unpredictable environment. Almost any host or device in the network can communicate with any other device at any time. This unpredictable nature requires more software support for verifying access permission, setting up sessions, and routing transactions to the correct software service.

This unstructured environment, which assumes that the attached devices are peers, is called *peer-to-peer.* Several workstations and mainframe computers can be connected together. Each is independent of the others, and occasionally they share information using one of the network protocols. Each workstation and mainframe is a peer to the others. Since this environment is similar to the way the telephone system operates where each telephone is a peer, analogies with the telephone system are common.

Networks are used for transferring not only data with error-free delivery, but also voice and recently video, where delivery on time is the main requirement, and error-free delivery is secondary. With video, for example, if delivery is late, the data is useless; but if we lose a pixel or two, we won't notice, as long as the picture doesn't flicker.

Supported protocols

Fibre Channel attempts to combine the best of these two opposing methods of communication into a new I/O interface that meets the needs of both *channel* users and *network* users.

Note that the word "Fibre" in the name "Fibre Channel" does not imply only the optical fiber media. With this spelling, the designers of this protocol use it to cover any serial media, such as copper coax or twisted pair, in addition to optical fiber.

Note also that the word "Channel" in the name "Fibre Channel" does not mean that it is only a *channel* protocol, as it might wrongly imply. Fibre

3

Channel defines a protocol which can be used for both channels and networks with equal efficiency.

Figure 1 shows the early support of various channel types and networks in Fibre Channel. This figure also shows that Fibre Channel supports the transmission of Asynchronous Transfer Mode (ATM), IEEE 802, and other network traffic. For those familiar with the Internet Protocol (IP), services such as electronic mail, file transfers, remote logins and others are supported by Fibre Channel at the new speeds.

These are important features when connecting Fibre Channel based systems to the main networks of the world and also to local networks already installed. This includes SONET-based ATM systems and local area networks (LANs), like Ethernet.

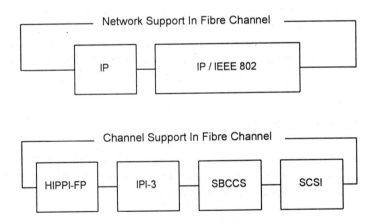

Figure 1. Supported Channels and Networks

A very important contribution of Fibre Channel is that these two interface types, channels and networks, can now share the same physical medium.

In the last few years, I/O channels have expanded to include network applications, such as using SCSI to interconnect two workstations. Similarly, networks move data between systems and file servers using network file transfer protocols, such as the Network File System (NFS).

With Fibre Channel, we now can use the same physical medium and the same physical transport protocol through a common hardware port to do both channel and network activity. One can send information to a network connected via Fibre Channel to the back panel of a workstation, and also use the SCSI-3 Fibre Channel protocols to talk internally to the local peripherals, as disks and tapes.

Nodes, Ports, and FC Links

Fibre Channel devices are called *nodes,* each of which has at least one *port* to provide access to the outside world: to other *ports* in other *nodes.* See Figure 2. A *node* may be a workstation, disk drive or disk array, a medical scanning instrument, etc. Each *node* can have more than one *port.* A *port* can be an adapter including a part of software in the host; sometimes it is difficult to say exactly where a *port* ends and the *node* begins. Some parts of software may be shared by several *ports.*

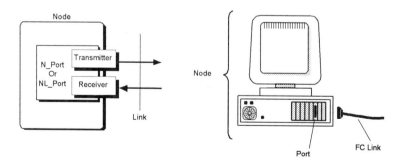

Figure 2. Node, Port, and a Link

There are several kinds of *ports*. Depending on their functionality, there are N_Ports, NL_Ports, F_Ports, FL_Ports, and E_Ports.

Each Fibre Channel port uses a pair of fibers: one to carry information into the port and one to carry information out. In Fibre Channel, fibers are electrical wires or optical strands. This pair of fibers is called a *link*.

The set of hardware components, such as media, connectors, and transceivers connecting two or more node ports, is called a *topology*. Each port in a node, called an N_Port in Figure 3, attaches to one of the topologies through a link. See Figure 3.

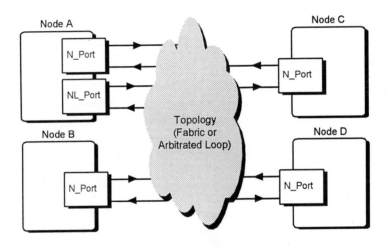

Figure 3. Topology

Do We Need Yet Another I/O Protocol?

Another I/O protocol! We already have many. Do we really need another one?

Yes, we do! It's true, today we have several channel protocols: SCSI, IDE, IPI-3, HIPPI, etc; we also have plenty of different network protocols: Ethernet, token ring, FDDI, etc. However, all these have serious handicaps holding up progress. Those protocols listed above represent only gradual evolution. This is because they carry with them the same old problems inherent in the design. The collision detection scheme in Ethernet means that at a certain loading level, the net will spend all its bandwidth in recovering from collisions, leaving no room for data transfer. The token ring scheme will always waste part of the bandwidth administering fairness - circulating the token. SCSI will always be limited by its topology because parts of its protocol use propagation delay values in the media to define timeouts, etc.

The Fibre Channel protocol is a new generation; it uses new concepts and bypasses all the limitations above. Let's summarize the new characteristics; we can expect that the new protocols will:

- *use serial media,* so that they can be easy and cheap to install and maintain. No more 50-pin connectors, thick rigid cables, etc.

- *be scalable*, to allow the introduction of newer technologies, such as optical fiber, laser light at higher speeds, and longer distances: kilometers rather than a few meters. Scalability should also mean that we will be allowed to transport many different protocols simultaneously. For example, Fibre Channel is capable of transporting many high level protocols, such as SCSI, HIPPI (channels), and TCP/IP, ATM (networks).

- *be asynchronous,* to allow maximum throughput, even when the system gets busy. Most protocols of today's generation start losing their efficiency under heavier loads, and some, like Ethernet, collapse entirely. Protocols using an asynchronous mode of transfer, like Fibre Channel, can maintain maximum throughput under maximum loading, or even during overload. The asynchronous mode functions like a heavily loaded highway where the cars are going at maximum speed, bumper-to-bumper, with all lanes in use. When another car wants to enter this highway, it simply has to wait for an available slot, nothing more; traffic on the highway is not affected by the number of cars waiting to enter. On the other hand, if we use the collision detection scheme of Ethernet, the waiting car would be allowed to crash in, and we would have to call the police, the tow truck, and the ambulance. Time would be spent recovering from the collision. A solution to congestion used in the token ring network is to allow only one car at a time on the highway. That way you guarantee that you don't crash, but you also guarantee that you waste part of the precious bandwidth.

- *be very, very fast*, so that when using lasers and optical fibers, we can transmit at enormous speeds. Fibre Channel at the maximum speed today of 1 Gbps can easily move 100 MBps on each fiber, even though each byte is sent as a string of 10 bits - 8 bits encoded to 10 bits - in serial fashion.

- *be switched*, to get away from shared media as with Ethernet, or shared bandwidth as with token ring and FDDI. The new high bandwidth networks use a switched Fabric to interconnect all participating nodes, providing many concurrent connections. There are a few other characteristics to be found in the new generation of communications protocols.

- *allow mixing media and speeds*, using single-mode optical cables for distances in kilometers, multimode cables for wiring within buildings, or using copper media (coax or twisted pair) for very short distances within enclosures.

- *separate the information transferred (payload) from the physical transport system*, allowing development and configuration of systems module by module. This capability leads to simplicity of the development or configuration process.

- *minimize time spent in error checking and error correction*, assuming that the optical media "never fails". Actually, recent measurements indicate that we may wait more than a day for a single bit error, therefore error checking and error correction is left to the higher level protocol only. This way a bigger part of the bandwidth can be used to transfer useful information.

Fibre Channel is a representative of this new generation of protocols. It is serial, scalable, asynchronous, extremely fast, and switched. It facilitates very fast and efficient communication today and allows for almost unlimited growth in the future. Read on for more detailed descriptions.

Fibre Channel protocol characteristics

Fibre Channel does not incorporate a command set as SCSI and IPI-3 do, but it does provide a mechanism to superimpose other protocols onto Fibre Channel. This is accomplished by using Fibre Channel as a carrier of those command sets in a way that allows the receiver to distinguish between them. This means that significant software investment in command sets for existing I/O interfaces, such as for device drivers and peripheral devices, will be directly transferred to Fibre Channel.

The separation of I/O operations from the physical I/O interface is another important feature of Fibre Channel. It allows using multiple command sets simultaneously. In Figure 1, you saw that various command sets, traditionally found with their own unique interfaces, are supported. Fibre Channel defines one common physical transfer mechanism for all these command sets.

Fibre Channel

- is a serial protocol
- is unaware of the content or meaning of the information being transferred
- increases connectivity among tens of devices, to thousands, or even hundreds of thousands of devices
- increases the allowable distance between devices
- increases throughput by four to five times over the most popular channels, and up to 100 times over popular networks

- uses asynchronous transfers when transporting information.

The following chapters will present more on how a network can be built using Fibre Channel.

INTERCONNECT TOPOLOGIES

Fibre Channel devices are called *nodes,* each of which has at least one *port* to provide access to the outside world. See Figure 2. The components that connect two or more ports are called a *topology.* See Figure 3. All Fibre Channel systems have these two elements: *nodes* with *ports*, and *topologies.*

Each Fibre Channel *port* uses a pair of fibers: one to carry information into the *port,* and one to carry information out of the *port.* In Fibre Channel, fibers are either electrical wires or optical strands. This pair of fibers, called a *link,* is part of each *topology.* See Figures 3 through 8.

Information is always transmitted in units called *frames* over these *links.* The Fibre Channel standard defines three *topologies:* Fabric, Point-to-Point, and Arbitrary Loop. The Fabric topology will be described first.

Fabric topology

A *Fabric* permits dynamic interconnections between *nodes* through *ports* connected to a *Fabric.* See Figure 4. Note that the word *Fabric* in this application can be seen as synonymous with the words *switch* or *router.* Each port in a node, called an *N_Port* or *NL_Port,* is attached to the Fabric through a *link.* Each port in a Fabric is called an *F_Port* or an *FL_Port.* It is possible for any node to communicate with any other node connected to other *F_Ports* of the same *Fabric,* using the services of the *Fabric.* In this type of topology, all routing of frames is performed by the *Fabric,* rather than by the *ports.*

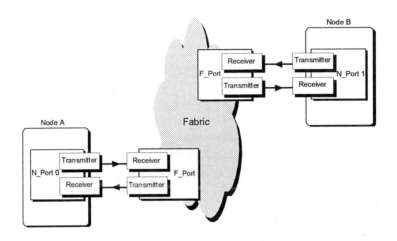

Figure 4. Two Nodes Connected Through a Fabric

This any-to-any connection service or *peer-to-peer* service is integral to the design of Fibre Channel. A system designed for *peer-to-peer* service can be used to emulate the host-type master-slave method of communication. So Fibre Channel can support channel and network protocols simultaneously.

Like a phone system

The function of the *Fabric* is similar to that of a telephone system: we dial a number, the telephone system finds a path to our destination, the phone rings, and the other party answers. At the same time, a third party may be calling a fourth party. Several calls are possible simultaneously. See Figure 5. If a switch or link goes down, the telephone company reroutes new calls on other paths and the callers seldom notice. Most of us are not aware of the intermediate connections that the telephone company switches make to complete our simple telephone call.

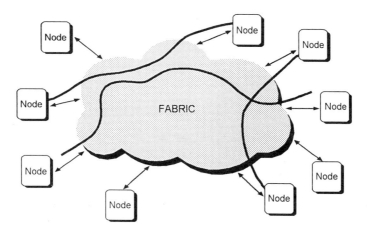

Figure 5. Nodes Connected Through a Fabric

We do give the phone company specific information about the call. For example, if the phone number starts with the digit 1 in the U.S., ten digits follow in the form of area code (3), exchange (3) and number (4). If there is no digit 1 to start the phone number, the call is within the caller's area code and we use only seven digits. This information helps the telephone company make the connection. The phone number is equivalent to the Fibre Channel *address identifier*. A part of the *address identifier* is used to determine the *Fabric domain*, corresponding to an area code, and the rest is used to identify the particular port.

Note that the telephone system is neither involved nor concerned with the contents of the conversation between the two calling parties; it merely provides the connection. Similarly, Fibre Channel provides the connection, and the superimposed protocols, such as SCSI or IP, carry the commands and packets. These protocols play a role similar to languages on the telephone system. We should view Fibre Channel and the other protocols as forming an integral part of the information exchange.

How to get from A to C

The complexity of a Fabric is similar to that of telephone switching devices; the equivalent Fabric devices are called *Fabric Elements.* In Figure 6 only one *Fabric Element* is shown, with four F_Ports labeled a, b, c, and d.

In Figure 6, if Node A needs to talk to Node C, the information is first sent to the Fabric at F_Port a. The Fabric makes an internal connection or set of connections to F_Port c. The information is then sent to Node C. It may be necessary to select several paths internal to the Fabric before arriving at F_Port c.

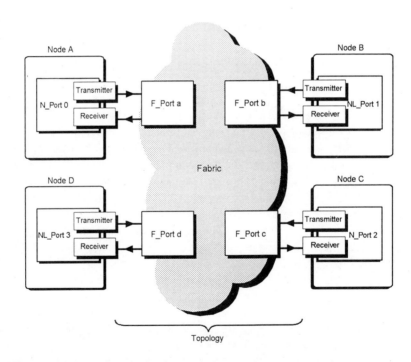

Figure 6. Fabric Topology With Four Nodes

The Fabric may consist of a single *Fabric Element,* or of several *Fabric Elements.* As in telephone systems, we don't know or care how many switches (telephone exchanges) we have to go through, as long as we are connected to the correct destination station.

One type of Fabric is called a *switched topology* or *cross-point switch topology.* Routing frames through the various switches is accomplished by having the Fabric Elements interpret the *destination address identifier* in the frame as it arrives at each Fabric Element.

As mentioned before, the Fabric can be implemented physically as a single Fabric Element with several F_Ports, as shown in Figure 4 or Figure 6, or as a set of several connected Fabric Elements. The routing or switching at each fabric element is transparent to the two N_Ports connected via F_Ports to the edge of the Fabric.

When the topology is separate from the nodes, as it is with the telephone system and Fibre Channel, new technologies may be introduced for the fibers within the Fabric. New speeds and new functions may be introduced into the Fabric without making all previous investment in the existing nodes obsolete. Fibre Channel allows a mix of attachments with different speeds or capabilities.

Point-to-Point topology

The Point-to-Point topology is the simplest one. Only two ports are used, connected by a link. See Figure 7. The transmitter of each port is connected directly to the receiver of the opposite port. There is no ambiguity in addressing, and there is no question of availability. Note that in this topology, characteristics of the two ports must be compatible, with the same speed, same media, same protocols, etc.

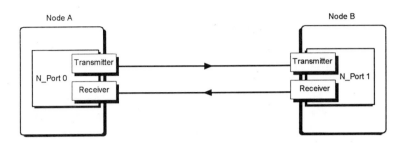

Figure 7. Point-To-Point Topology

Arbitrated Loop topology

The Arbitrated Loop (AL) topology is more involved but less complicated than a Fabric. It can have up to 127 ports, all connected in a serial (loop) fashion. See Figure 8. All ports must have compatible characteristics, as to speed, medium, etc. An advantage of this topology is its lower cost compared to the Fabric. A disadvantage is in that the bandwidth is shared: only two ports can be active at one time; all other ports on the loop must act as repeaters. The Arbitrated Loop topology is used in systems where low cost is important, and where sharing the bandwidth of a link is acceptable.

The ports in an *Arbitrated Loop* topology, called *NL_Ports* and *FL_Ports,* are slightly different from N_Ports and F_Ports. They contain all of the functions of N_Ports and F_Ports and can operate correctly with a Fabric. The "L" in the port name indicates a port that handles the Arbitrated Loop protocol.

In the case of an *Arbitrated Loop,* as in a Token Ring protocol, each port sees all messages, and passes and ignores those not addressed to that port. The *Arbitrated Loop* protocol provides a token acquisition protocol.

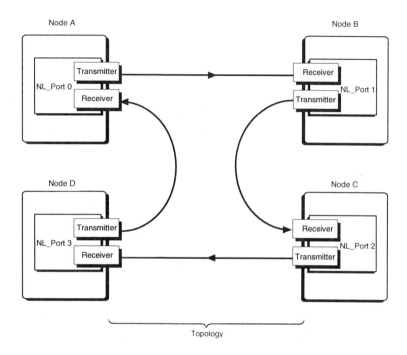

Figure 8. Arbitrated Loop topology with four nodes

In many installations, the Point-to-Point, Arbitrated Loop, and Fabric topologies will be combined. For example, a group of six clerks may be connected using the Arbitrated Loop topology. One of the ports on the loop - the FL_Port - is then connected to a Fabric switch, attaching the six clerks to a switched network for the rest in the company. Another example is using the Arbitrated Loop topology to connect local peripherals to a workstation, and using a second N_Port on the same node to connect to the Fabric. A disk array may be attached to the same workstation, using the Point-to-Point topology, to provide guaranteed performance characteristics for accessing that data.

17

More on topologies

To better understand the function of a Fabric topology, let's continue the analogy with the telephone system: we dial the phone number for a friend. We don't have to know the precise route through the telephone system to our friend's house when we call. The routing is taken care of by the telephone system. The Fabric of Fibre Channel does the same function - you give a destination address and the Fabric routes frames to that destination N_Port.

If you dial a number incorrectly, the telephone company tells you the number is not in service. The Fabric rejects frames for invalid destinations in much the same manner.

Just as the telephone company can and does configure many routes between various points to provide reliable service, a Fabric can have multiple paths between *Fabric Elements* to carry more reliable traffic. This also permits backup paths in case of an element or link failure.

The Fabric and Arbitrated Loop topologies of Fibre Channel may be intermixed in one system to provide a variety of service and performance levels to the nodes. In addition, a Fabric may use other networks such as ATM over SONET between Fabric Elements, to achieve longer distances between nodes than can be achieved by the links defined for Fibre Channel. These special links would be found between Fabric Elements distributed over a large geographic area and not directly connected to nodes.

The capability to add other types of links between Fabric Elements, called expansion ports or E_Ports, enhances the value of any device attached to the Fabric. Certain attributes of Fibre Channel and a Fabric permit ports of various speeds and media types to communicate with each other over either short or long distances when the Fabric is present. The Fabric can also add technology improvements within the Fabric, without altering the N_Ports in any way. Most of the benefit of the new technology is indirectly passed on to the nodes because of the improved speed, reliability, or distance of communication within the Fabric.

How many Ports can there be?

The Fabric is constrained only by the number of N_Ports that can be identified in the destination address field in the header of a *frame*. That limit is set at a little over 16 million ports that can be concurrently logged in to a Fabric with the 24-bit *address identifier*. For single integrated systems, that should meet all needs for quite a while.

In the Fabric topology, the *address identifier* is divided into three parts: *domain* (8 bits), *area* (8 bits), and *port* (8 bits), for a total of 24 bits. These parts are similar to the phone number components of area code, exchange, and number.

Only 127 ports are allowed in the Arbitrated Loop topology on a single loop. We find that in practical implementations the number is actually much lower. The Arbitrated Loop uses the lowest byte of the address identifier for addressing within the loop. The higher bytes are reserved for addressing the external network connected by the Fabric.

The port limit for the Point-to-Point topology is two ports connected by a single link.

FUNCTIONAL LEVELS

Some topics just naturally belong together and so it is with Fibre Channel. Items dealing with producing reliable and testable optical links are quite removed from deciding how to recover from a missing frame. These separate areas of interest are called functional levels in the Fibre Channel standard. There are five levels defined; each is labeled FC-x.

FC-0   defines the *physical* portions of Fibre Channel, including the media types, connectors, and the electrical and optical characteristics needed to connect ports. This level is in the FC-PH standard.

FC-1   defines the *transmission protocol*, including the 8B/10B encoding, order of word transmission, and error detection. This level is in the FC-PH standard.

FC-2   defines the *signaling and framing protocol*, including frame layout, frame header content, and rules for use. It also contains independent protocols such as login. This is the bulk of the FC-PH standard.

FC-3   defines *common services* that may be available across multiple ports in a node. This level has no standard now.

FC-4   defines the *mapping* between the lower levels of Fibre Channel, and the command sets that use Fibre Channel. Here you will find SCSI, IPI-3, HIPPI, etc. Each of these command sets is given a separate standard, so that those interested get only what they want. If you are doing SCSI-3, you probably don't care about IPI-3 or HIPPI.

Figure 9 shows the internal structure of a Fibre Channel node with one N_Port. Multiple N_Ports per node are allowed. If a node has more than one N_Port, the levels FC-0 and FC-2 are replicated for each N_Port. Levels FC-3 and FC-4 are common across multiple N_Ports.

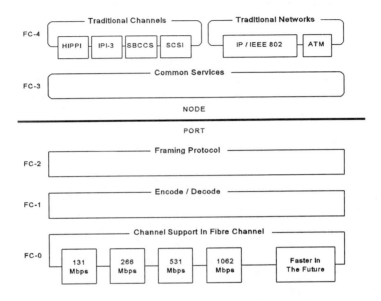

Figure 9. Fibre Channel Node Structure

Each of the functional levels identified above is shown in Figure 9. The Common Services layer has no defined components at this time. However, as the development work on the standard continues, new functions will be added at this level.

At the FC-4 level, a node may not implement all of the various options allowed by the standard. Each node may implement one or more of these protocol mappings. At the FC-0 level, only one speed and media choice, as presented later, may be implemented in one N_Port. However, each different port on a node may implement a different combination of speed and media. The FC-2 level also supports many options from which an implementor must choose. There is some work being done by industry groups to define *profiles* that specify the operating environment required for some applications (e.g., SCSI-3/FCP, HIPPI, IP, etc). There are no options at the FC-1 level.

Fibre Channel profiles are documents that specify a subset of the options in a standard to be used for certain applications. Profiles exist for SCSI-3 FCP, IP on Fibre Channel, disks on the Arbitrated Loop, etc. The purpose of a profile is to help various devices, manufactured by different vendors, interoperate while performing the same applications. Devices are more likely to interoperate if they conform to a profile.

Several of the FC-4s currently being defined are briefly described below. Each FC-4 is being developed as an individual part of the standard in its own separate document. An implementor can concentrate on just the FC-4 or FC-4s of interest and not be bothered with the requirements for the others.

However, if expansion is a possibility in your product development or purchasing plans, then the other candidate FC-4s should be checked so that a superset of the port requirements can be in place when the expansion occurs. For example, a system vendor putting the IPI-3 protocol in and later expecting to add the Internet Protocol (IP) should carefully consider both FC-4s before choosing a port design, since the requirements are different.

Protocols transported by Fibre Channel (FC-4)

The following sections address the channel and network protocols that will or can use Fibre Channel for their transport system. The original higher level command sets and functions will be retained, while the lower level will be replaced by Fibre Channel.

As an example, SCSI will retain its command set and a big part of the device drivers and peripheral device code, but instead of using the traditional 50-wire or 68-wire cables, connectors, and SCSI protocol chips, this lower level will be handled by Fibre Channel. The host adapter driver, rather than talking to the parallel bus SCSI chip, will assemble packets and will hand them over to the highest level (FC-4) of the Fibre Channel protocol driver. Similarly, on the other end of the FC link, the receiving FC-4 module will pass the received packet to the receiving peripheral device driver.

Today, the protocols in the sections below can be mapped to the Fibre Channel through its FC-4 level.

HIPPI   *High Performance Parallel Interface.* One of the original goals of Fibre Channel was to replace HIPPI with a high-speed serial interface. HIPPI is a full-duplex protocol: its main goal is to ship large quantities of data with little or no overhead. HIPPI has no native command set. It has only *sources* and *destinations*; you will usually find IPI-3 used when a command set is needed. HIPPI runs at 100 MBps in both directions on short parallel cables.

IPI-3   *Intelligent Peripheral Interface-3.* IPI-3 permits data transfer over long distances, since it has no timing requirements during data transfer. IPI-3, in its parallel version, is a half-duplex protocol. In its Fibre Channel version, it retains that half-duplex nature for an operation. IPI-3 specifies speeds up to 100 MBps on short parallel copper cables. It supports multiple device types.

SBCCS   *Single Byte Command Code Set.* This protocol, as its name implies, transfers commands with a single byte of information. It is a half-duplex protocol in parallel busses and retains much of that character in the Fibre Channel implementation. SBCCS is derived from the IBM byte/block multiplexer channel, originating with IBM systems in the 1970s. The IBM proprietary version, called ESCON, uses fiber optic links and 8B/10B encoding at 20 MBps in each direction.

24

A project is underway in the X3T11 ANSI Committee to convert ESCON to a standard, called SBCON. SBCCS is its follow-on with Fibre Channel.

SCSI   *Small Computer System Interface.* SCSI is a very popular interface for peripherals in workstations and larger PCs. It is a half-duplex interface in its parallel versions. The data rates run from 5 to 40 MBps. SCSI defines commands, status, and sense data for each device class it supports. The Fibre Channel implementation has two versions: Fibre Channel Protocol (FCP) and Generic Packetized Protocol (GPP).

The Fibre Channel Protocol (FCP) retains the half-duplex nature of parallel SCSI within a task. FCP has been selected by the Fibre Channel Systems Initiative group (FCSI) as their principal implementation of SCSI on Fibre Channel.

The Generic Packetized Protocol uses a full-duplex model to operate efficiently over long distances. It has optional features to allow more than one port pair to be used for the transfer of information, and an automatic failure mode if one of these port pairs becomes inoperable.

IP   *Internet Protocol.* The Internet Protocol (IP) is part of the suite of protocols that comprise Transmission Control Protocol/Internet Protocol (TCP/IP). Internet is one of the most successful networking protocols. It was one of the first where the designers separated function from form: they separated the transferred information from the method of transfer. This protocol is speed independent.

IEEE 802   The Institute of Electrical and Electronics Engineers (IEEE), which is also part of ANSI, developed the *IEEE 802 standard*, written in several parts and generally concerned with *Local Area Networks (LANs)*. In general, the data rate in LANs is from 1 to 10 MBps. Some parts are very specific about a particular physical layer. Other parts specify the form of information transfer independent of the physical layer.

The various parts of the standard are identified as IEEE 802.x, where x is the part number. IEEE 802.2 is the *Logical Link Control (LLC)* part. This is the part that Fibre Channel references. Fibre Channel becomes a new physical layer or *Medium Access Control (MAC)* layer in addition to the four it already defines. Another existing MAC layer is the *Fibre Distributed Data Interface (FDDI)*.

Applications that use Ethernet, token ring, FDDI, or token bus today can switch to Fibre Channel by changing only the MAC layer in their systems, or by just adding a new one. IEEE 802 does not specify a command set. Cooperating peers select a protocol and use IEEE 802 to transfer information. IEEE 802 is an integral part of the Internet protocol as well, where it allows small work groups to participate in the wider network.

ATM   *Asynchronous Transfer Mode*. This is a recent and widely referenced telecommunications protocol. Like Fibre Channel, ATM is a link layer protocol. ATM is related to the *Integrated Services Digital Network (ISDN)*, which includes broadband ISDN and ATM. ATM uses T1 services or *SONET* as a physical layer. These and other services form the basis for *Metropolitan Area Networks (MANs)* and *Wide Area Networks (WANs)*.

Fibre Channel activity will concentrate on the *ATM-Adaptation Layer 5 (AAL 5)*, which supports computing environments rather than voice transfers. By providing an interface at this level, several physical layers can be used within the structure of ATM, much as IEEE 802 has multiple physical layers. ATM provides no command sets, as is common with networking protocols. The peer devices again decide on an internal protocol for the information content, and the physical layer neither knows nor needs to know what is being communicated.

CLASSES OF SERVICE

With humans, there are various kinds of communication strategies. Sometimes we need a direct face-to-face meeting; sometimes we can send a fax or electronic mail to one another or have short telephone conversations; sometimes, as with advertising, we just broadcast information in the hope that the message gets through. Each communication strategy meets a different need and all humans use one kind or another at various times. The Fabric topology supports similar communication strategies, but it has different names for them.

The Fibre Channel names for these communication strategies are gathered under the title of *classes of service.* Figure 10 summarizes the *classes of service* in Fibre Channel.

Class of Service	Description
Class 1	"Connection Oriented" or dedicated connection
Class 2	"Connectionless" operation
Class 3	"Datagram" service
Class 4	Fractional bandwidth (voice & video), coming in FC-PH-2

Figure 10. Classes of Service

Class 1 provides a "*connection oriented*" mode of operation, similar to circuit switching in a telephone system operation. The caller dials the destination and a dedicated circuit is established. After this, dialog starts, with or without pauses. Frames require confirmation of receipt. At the end of the conversation the parties disconnect. As can be seen, this class of service has three phases: 1) setting up the connection; 2) transferring the information; and 3) closing down the connection. An advantage of Class 1 is dedicated connection; its disadvantage is that usually nobody can access the calling parties while the connection is in operation.

Class 2 provides a *"connectionless"* more interactive mode of operation, also called frame switching. No dedicated connection between the two communication parties is set up. This class of service allows a stream of frames to be sent to different destinations quickly. Class 2 also requires frame confirmations by the recipient.

Class 3 is also a *"connectionless service"*, sometimes called "datagram". It is a type of transmission which does not require confirmation of frame delivery.

Class 4 is a *"connection oriented "* mode of operation, but it uses *virtual connections* rather than *dedicated connections*. Class 4 distributes the bandwidth of a port among several destinations. This guarantees access to those destinations, but without allocating all of the bandwidth as with Class 1. This class of service is being defined in FC-PH-2.

In future versions of Fibre Channel more classes of service are being defined. One is the *"isochronous service"* for voice and video; another is *"Buffered Class 1"*.

CHOOSING CABLES & TRANSCEIVERS

Early in the design cycle, a system designer has to make several choices. First, defining the general physical layout and specifying all point-to-point distances. Then comes the choice of individual devices and cabling. The designer may also start from the other end and chose the devices, then based on those, calculate the maximum lengths of individual FC links.

This chapter addresses the media choices, speeds available, and other terminology used in laying out a Fibre Channel network.

Designers should be familiar with the following vocabulary:

Distance

We choose between the copper and the optical fiber based on the required length of each link. We also must choose between different types of fiber - the single-mode or multimode. The media type and speed usually determines the maximum length of a link. A workstation in a work group may use a copper link over short distances, multimode fibers between floors in a building, and single- mode fibers for long distances between buildings.

Optical fiber or copper cable

Each link is driven optically or electrically. The optical and electrical types can be combined in a single system when there is a Fabric or other media-type converter available. A Fabric is the only speed converter available in Fibre Channel.

The optical fiber menu is shown in Figure 11. The unit of measure for distance is kilometers.

Laser Type	Effective Rate MBps	Distance kilometers	Baud Rate Mbaud
9-Micron **Single Mode Fiber**			
Shortwave	100	up to 10	1062.5
Long wave	25	up to 10	265.6
50-Micron **Multimode Fiber**			
Shortwave	100	up to 0.5	1062.5
Shortwave	25	up to 2	265.6
Long wave LED	25	up to 2	265.6
62.5-Micron **Multimode Fiber**			
Shortwave	100	up to 0.175	1062.5
Shortwave	25	up to 0.7	265.6
Long wave LED	25	up to 1.5	265.6

Figure 11.   Optical Fiber Menu

The electrical fiber menu is shown in Figure 12. In this figure, the unit of measure for distance is changed from kilometers to meters.

Type	Effective Rate MBps	Distance Meters	Baud Rate Mbaud
Video Coax Fiber			
ECL	100	up to 25	1062.5
ECL	25	up to 75	265.6
Miniature Coax Fiber			
ECL	100	up to 10	1062.5
ECL	25	up to 25	265.6
Shielded Twisted Pair			
ECL	25	up to 50	265.6

Figure 12. Electrical Fiber Menu

The optical cable

The active part of the optical fiber is constructed out of the core (the optical conduit), surrounded by cladding (to keep the light in the core), and fiber coating wrapped around the cladding.

The optical fiber is very thin and could be damaged easily. For that reason, it is packaged with protective coating and mechanical buffering, then all is enclosed in a protective jacket.

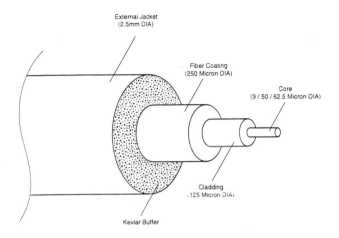

External Jacket
(2.5mm DIA)

Fiber Coating
(250 Micron DIA)

Core
(9 / 50 / 62.5 Micron DIA)

Cladding
(125 Micron DIA)

Kevlar Buffer

Figure 13. Optical fiber cable construction

The fiber core itself is 9 micron (1 micron = one millionth of 1 meter) in diameter for Single-mode, or 50 or 62.5 micron for Multimode cables. For comparison, a human hair is about 75 micron thick.

Fiber types: single-mode or multimode

Light travels through the core of the fiber, and the cladding keeps the light within the core by reflecting photons back into the core. Depending on several factors, we use a single-mode (a single stream of photons) or multiple modes (multiple streams of photons). In a way, we are talking about focusing the light beam into the fibre, and keeping it focussed while it travels through the fiber.

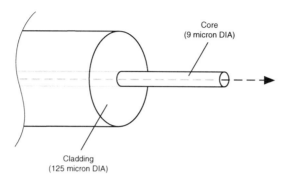

Figure 14. Single-mode fiber

In Figure 14 note that the single-mode components require better focusing on the source side and finer optical fiber (9 micron), resulting in higher cost. On the receiving end we get only a single ray (at least theoretically) with minimum dispersion. Single-mode cables are used for longer distances.

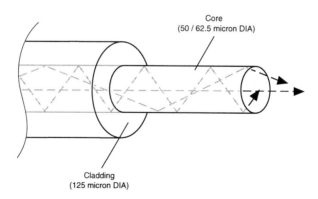

Figure 15. Multimode fiber

33

The multimode components are cheaper because they don't require such expensive optics and such fine fiber. However, the quality of the signal at the receiving end is lower. The multimode cables are used for shorter distances (to 1 or 2 kilometers or less) where the low cost is important.

For illustration, we can compare the single-mode as shooting from a rifle, and the multimode as using a shotgun. See Figures 14 and 15.

Wavelength: shortwave or long wave

The Fibre Channel is a broadband protocol. This means that it uses a carrier frequency (always present) modulated by the signal frequency. The wavelength is the term to define the "color" of the light used as the carrier. It is given in nanometers (nm); one nm = one billionth of a meter. The wavelength of visible light is in the range approximately between 490 nm and 750 nm.

The 780 nm and 850 nm wavelengths are called shortwave; the 1300 nm and 1550 nm wavelengths are called long wave. The shortwave signals can use multimode fiber only, the 1300 nm can use either multimode or single-mode fiber, and the 1550 nm can use single-mode only.

Transceivers use transmitters with lasers or LEDS

A transceiver has both a transmitter and a receiver. The transmitters with lasers use a monochromatic single-mode source for high power and for long distance (single-mode fiber has less loss), or LEDs with a multimode source for shorter distances and if low cost is important.

Connectors

The Fibre Channel standard specifies the SC type connector for optical fibers, and the DB-9 or BNC connectors for copper.

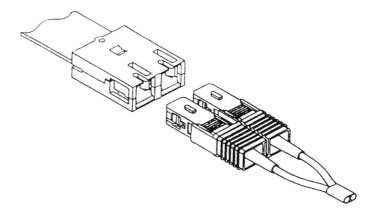

Figure 16. SC connector used for optical fiber

Duplex cables and connectors are used for Point-to-Point or Switched topologies and FC-AL hubs, and simplex cables for the Arbitrated Loop topology. You also need to specify the single-mode, multimode (50 or 62.5 micron), or copper medium type.

Open Fibre Control (OFC)

The Open Fibre Control (OFC) circuitry is used for safety when using higher power transmitters and shortwave lasers. When the receiver detects loss of light, as with a disconnected cable, it shuts down the transmitter in the same transceiver module. The OFC is required on all medium power and high power shortwave 780 nm lasers, and on higher power shortwave 850 nm lasers. It may also be required on some higher power long wave lasers. If the OFC is used, it must be installed on both ends of the same FC link, to allow it to start up the link.

Power budget

Laying out the FC network requires that power levels be respected. The designer needs to know the output power of the transmitter in dBm on one end of the link, and the sensitivity of the receiver in dBm at the other end. Subtracting the two values gives the power budget, in dB, available for the link. Then the sum of the losses in connectors, splices, the fiber itself, and the required margin can be compared to the power available.

Mix and match

Speed conversion is important to system integrators. A mix-and-match system is quite reasonable and usually less expensive than a homogeneous single speed system. There is no need to specify that a system with a Fabric has all ports with the same type of media, lasers, distance, or speed capabilities. The performance capabilities allow the implementor to meet a wide range of performance, cost, and distance requirements. A Fabric is normally required for larger systems. It is possible that Fabrics will be designed so that F_Ports and FL_Ports can be changed easily from one type to another with little or no interruption to the rest of the Fabric.

PACKETS and FRAMES

In network jargon, a block of information to be processsed or transferred is called a *packet.* When other information is added for control, and the packet is placed on the physical wire, the packet is sent as a sequence of one or more frames. A *frame* is a block of bytes together with control information. Since the packet is constructed without knowledge of the transport system, it may be larger than can fit in one frame. Thus, a packet is sent as a sequence of one or more frames.

In Fibre Channel, it's a little more complex. The *frames* are grouped into *sequences*, and *sequences* are grouped into *exchanges.*

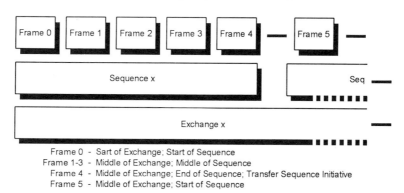

Frame 0 - Sart of Exchange; Start of Sequence
Frame 1-3 - Middle of Exchange; Middle of Sequence
Frame 4 - Middle of Exchange; End of Sequence; Transfer Sequence Initiative
Frame 5 - Middle of Exchange; Start of Sequence

Figure 17. Frames, sequences, and exchanges

For example, when an application needs to move a 100 MB file, it will hand the request to the operating system (OS) as one *packet.* The Fibre Channel port driver in the OS will package this request as a simple exchange. Before the actual transfer, the driver will divide the exchange into several *sequences* (for example, 20 KB long each), and each sequence into several *frames*: in our example, ten 2 KB frames. Error checking and reassembly is handled by the sending and receiving ports, frame by frame, sequence by sequence.

The Fibre Channel *frame* format is shown in Figure 18.

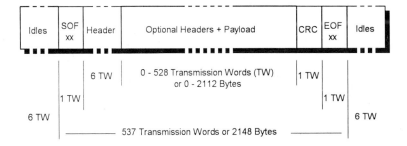

Figure 18. Format of a Fibre Channel Frame

Fibre Channel frames consist of multiples of *Transmission Words* (TW). One transmission word is four bytes long.

The length of a frame can be up to 2148 bytes, out of which the maximum *payload* can be 2112 bytes. Without the *payload*, the shortest Fibre Channel frame is 36 bytes long. A full *payload* is 2048 bytes of data together with 64 bytes for optional headers.

As an example, we could transfer a SCSI command consisting of the Command Descriptor Block (CDB), data, and message in the same frame. Another example may be encapsulating a full Ethernet frame of 1512 bytes in a single Fibre Channel frame.

The *frame header* is used by the FC-2 transport level. The information to be transferred is placed in the payload, and the rest of the frame is used for transport control.

The *frame header* is divided into fields to carry the control information. See Figure 19.

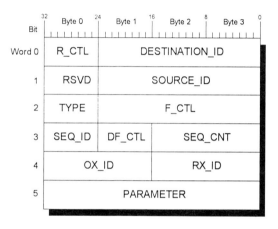

Figure 19. Fibre Channel Frame Header

In the *frame header*, the sender identifies itself, the destination, the protocol, and the type of information, etc. Some parts of the frame header are used by the Fabric for routing while the rest is used by the receiving port.

DATA - VOICE - VIDEO

Information transferred over networks or channels calls for different treatment in respect to its delivery requirements. We can divide the information into three groups:

- data
- voice
- video

When transporting *data*, the main requirement is reliability. Data may arrive at the destination a little late, but it must arrive **error-free.** In network terminology this is called a "*reliable*" or "*lossless*" type of delivery.

Voice and *video* are different. Both must be delivered **on time.** A little error or a lost frame here or there is usually not a serious problem. For example, in real-time video, the viewer may not notice a missing pixel in a picture or even a missing frame, one out of 30 per second. However a frame received out of sequence will hurt the quality of the picture.

A similar situation exists for *voice* transmission: a short missing segment in the sound will not degrade it noticeably or damage its understandability, but a segment of voice at the wrong time certainly will.

For this reason, protocols for *voice* and *video* must suppress out-of-order delivery frames and arrive on time. In network terminology, we call these "*non-reliable*" or "*lossy*" protocols.

Using these classifications, we can describe the transported information as follows:

1. DATA (computers, fax machines, ...) using *error-free, reliable,* or *lossless* protocols

2. VOICE (telephone) using *on-time, in order, non-reliable* or *lossy* protocols

3. VIDEO (multimedia, ...) using *on-time, in order, non-reliable* or *lossy* protocols

Fibre Channel has been designed primarily for data transfer. "*Reliable*" or "*lossless*" operating modes are defined for Classes 1, 2, and 4. However, Fibre Channel can be used for "*lossy*" voice or video transmission when using Class 3.

The ATM protocol was designed by WAN architects (the "telephone people") to handle primarily voice, and video in the future. It is a "lossy" protocol by design. When ATM gained recognition and started being considered for LAN and backbone applications, it was discovered that almost all information transported on today's LANs is the type of data that cannot use "*lossy*" protocols.

The quickest and simplest solution for ATM was to use higher bandwidth than the maximum expected, to prevent congestion, which would result in a loss of data. This solution would work, but it is obviously wasteful. A better solution is to develop a signaling mechanism, to work in a way similar to the way Fibre Channel works. This development effort is in process.

We can apply the criteria above to the Fibre Channel and ATM protocols. We can see that Fibre Channel is a protocol designed primarily for *data* delivery (Classes 1, 2, and 4), and secondarily for *voice* and *video* (Class 3). On the other hand, we can see that ATM is primarily a long distance Wide

Area Network (WAN) protocol, designed mainly for *voice* and *video*. In the near future it will also handle *data* in an acceptable way, although not as efficiently as Fibre Channel. The relationship is shown graphically in Figure 20.

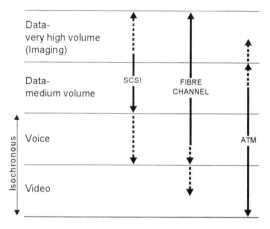

Figure 20. Fibre Channel vs. ATM - Effectiveness

We also can compare the Fibre Channel and ATM protocols as a choice for use in the WAN, LAN, or Channel applications, as shown in Figure 21.

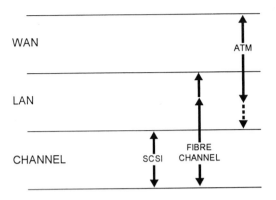

Figure 21. Fibre Channel vs. ATM - application

In conclusion, note that Fibre Channel and ATM have different characteristics, and that they should be viewed not as rivals, but rather as two complementary protocols. See Figure 22.

Fibre Channel is optimized for high performance data delivery, but also is equipped for handling voice and video reasonably efficiently.

ATM is optimized for voice and video, but also is equipped for reasonable performance in data delivery.

When designing a network (LAN or WAN), the architects should consider all these aspects, and decide which features and capabilities are important. As a result, we can expect Fibre Channel and ATM to be used in combination. We will see ATM traffic pass over Fibre Channel lines, and Fibre Channel traffic pass over the ATM lines.

We can expect that network fabric switches will be built with a combination of ports: some for Fibre Channel, and some for ATM.

The Fibre Channel-to-ATM protocol translation will be done in the Fabric itself, and the nodes will not be involved.

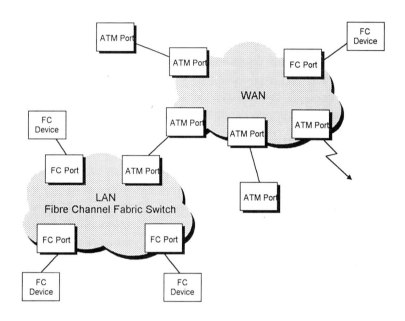

Figure 22. Fibre Channel and ATM are complementary

PERFORMANCE

A basic premise of Fibre Channel is that it can replace HIPPI. Although the parallel version of HIPPI does not allow long distance use, it is fast - 100 MBps. IPI-3 at 25 MBps is a second performance mark for Fibre Channel. These two speeds, and 50 and 12.5 MBps, make up the four speed choices for Fibre Channel: 100, 50, 25, and 12.5 MBps. To support a net rate of 100 MBps, we must have a bandwidth of ((800 + 200) + 62.5) or 1062.5 megabaud (Mbps) to cover the data encoding and framing protocol overhead. The term baud generally means bits per second.

Since the calculation is linear, we just divide by 2, 4 or 8 to arrive at the corresponding rates for the lower speeds. The complete set of encoded bit rates for Fibre Channel is 1062.5, 531.25, 265.625, and 132.8125 megabaud. Figure 23 shows a rate comparison between the encoded bit rates and the maximum effective data rates on a single fiber. Remember that Fibre Channel is a full-duplex system, so that the effective transfer rate may be twice as high, if sufficient data can be generated in both directions.

Encoded Bit Rate - Mbps	Maximum Data Rate - 1 Fiber - MBps
1062.5	100
531.25	50
265.625	25
132.8125	12.5

Figure 23. Fiber Bit Rate vs. Maximum Data Rate

The low overhead of Fibre Channel is possible partly because of its hardware intensive requirements. Each port must keep its link alive by continuously sending frames or *Idles* on its outbound fiber. Each port in a system receives a continuous stream of frames and *Idles* at full speed - 1.0625 gigabaud. There is no time to process the incoming signal or to make other decisions in any type of processor. There must be a real-time hardware assist in each port if the kind of data rate advertised is to be achieved.

Overhead is a fixed or proportional cost of transmission that cannot be eliminated. In Fibre Channel, the overhead is measured in lost time (increased latency) or bytes per second of transfer rate lost. The *8B/10B encoding* is the largest component of overhead at 25%, and the increased baud rate compensates for it. Also the increased baud rate compensates for the framing protocol header and trailer. The effective transfer rate in Mbps is close to 75% of raw transmission speed. That is, an increase in electronics and optics speeds achieves the desired net transfer rate.

When HIPPI uses Class 1 or Class 3, it can get more than 103 MBps over a single fiber. If HIPPI uses Class 2, the effective transfer rate is a little over 100 MBps, but Fibre Channel meets the requirement stated earlier in any class of service.

The capability to sustain such data rates depends on the node and port designs. Fibre Channel makes it available. The challenge now is for the systems and devices to figure out how to make it happen.

Users sometimes compare Fibre Channel to ATM/SONET. At a closer look however, one can see that these two protocols have different strengths and should be viewed as complementary.

The graph in Figure 24 shows a comparison of transfer speeds for Fibre Channel and ATM/SONET.

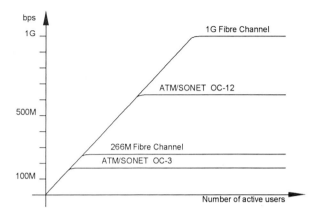

Figure 24. Fibre Channel vs. ATM / SONET - Transfer Rate

TESTING

Fibre Channel is a revolution in the channel and network industries, since it combines both protocols in the same transport layer. Fibre Channel brings with it a new set of testing challenges. Figures 3 through 8 show the magnitude of the problem.

Fibre Channel testing requires a completely new set of equipment and procedures. The physical layer, the framing protocol, and the logical layer mapping (FC-4 level) are all new. Also each of the topologies is new.

The testing challenges fall into four areas, roughly aligned with the levels of Fibre Channel and the topologies.

- The physical layer (FC-0 and FC-1) requires testing the transmission network.

- The framing protocol layer (FC-2) requires testing the content of individual frames and the overall content across many frames.

- The protocol layer (FC-4) requires testing one or more FC-4s per node and the interaction between them.

- Finally, the topologies require independent testing, with the Fabric topology being the most difficult to test.

One type of tester needs to be capable of generating, receiving, filtering, and analyzing Fibre Channel frames. A second type of tester monitors frame traffic flowing on a link, and filters information selected by the tester from the bidirectional stream of frames on a link. Fibre Channel will also be used as a network; therefore real-time monitoring, statistics gathering, and management functions will be needed. For this purpose we will need a monitoring instrument known as a "sniffer", currently used on today's networks.

The Fabric topology presents a big challenge, since there can be many links to monitor simultaneously, and much information to collect into an understandable report. In this case, each link of interest needs to be monitored.

When testing the Fabric, we need to exercise all switching, routing, and internal Fabric queuing functions. We need to measure internal latency under different traffic patterns, and under varying loads, including congestion. Proper testing of the Fabric is extremely important because the Fabric is positioned in the middle of the network and its failure can bring the entire network down.

A combination of test generators and link monitors is needed to test the physical system completely. End-to-end testing provides a means to test the interaction of independent nodes, once the topology has been qualified.

We need to have a functional link before we can inject and record frame traffic. So the first level of testing should include sending patterns of data (not frames) and performing *bit-error-rate tests* (BERT).

Jitter on optical or electrical fibers is a fact of life in Fibre Channel and similar networks. It is caused by several factors; one of them is the use of several independent clocks along the segmented paths. We should also test for resilience of the receiver ports and their tolerance of jitter. On the receiving end of the FC links we also need to verify that jitter generated by the transmitter and in the link is within tolerance limits.

BERT and jitter tests indicate the overall stability of each link type. This level of testing qualifies the fibers, connectors, transmitters, and receivers.

Once the physical layer is stable, tests begins for the other three categories that require independent testing. The University of New Hampshire Interoperability Lab (Fibre Channel Consortium) is currently planning to provide third-party testing service on these three categories. Work is under way to develop the testing methodology.

Thorough testing by a third party will greatly reduce the interoperability difficulties. Products that have only self-testing or company internal testing before coming to market frequently have those problems. Other channel-type interfaces lack this necessary service. There are some certified testing centers in the networking world now, and they may help by expanding the available services.

THE FUTURE OF FIBRE CHANNEL

Fibre Channel is here and is receiving much attention from most high-tech corporations. Some components that were available in early 1994 are making systems a reality. Since the time to produce a new product is considerable, especially with new technologies and interfaces, the plans to produce these products have been in place for several years. All that planning and development has resulted in new products now available.

All the components to build a Fibre Channel system are available now. Some of these components were built based on an earlier level of the standard, but valuable development work was achieved in 1993 with these. These systems can operate at either 25 or 100 MBps. There were baseline feasibility systems, or beta test systems, as they are sometimes called.

There are many suppliers of Fibre Channel components. Many high-tech corporations have parts, are buying parts, or are planning to buy parts for Fibre Channel.

The first beta testing site for Fibre Channel is at Lawrence Livermore National Laboratory in California. The laboratory has 32 by 32 cross-point switches, nodes, ports, and software running on several UNIX platforms.

The Fibre Channel consortium of the University of New Hampshire (UNH) InterOperabiltity Lab is developing test suites, for testing Fibre Channel devices for conformance to ANSI standards.

The ANSI committee X3T11, formerly X3T9.3, is responsible for developing the suite of Fibre Channel standards. Open meetings, with representatives from many companies, are held at regular intervals. This committee produces the standard, but it is not responsible for making judgments about implementations, or for compliance testing of them. Compliance or conformance testing is usually performed by independent companies.

The Fibre Channel Association (FCA) is a cooperative organization also independent of the ANSI committee. Membership is open to any company for a fee. Its purpose is to promote the use of Fibre Channel.

The Fibre Channel Systems Initiative (FCSI) was made up of representatives from Hewlett-Packard Company, IBM Corporation, and Sun Microsystems Corporation. Its purpose was to promote early interoperable implementations of Fibre Channel for their workstation products. By defining interoperable subsets of the standard, products from the three companies can work together. Also, suppliers of Fibre Channel products to these companies will be required to meet the subset requirements. The FCSI fulfilled its mission and has been dissolved.

Much work remains to be done on extensions to Fibre Channel. Two imminent features are faster link speeds, and a new class of service that simulates a just-in-time-data-arrival service, called Class 4. Also the concept of a hunt group has been added, where a node defines a collection of N_Ports to be equivalent for the purposes of sending and receiving frames. This has the potential to improve performance, and possibly to allow using multiple port pairs to increase transfer rates further. These are just a few of the new things on the horizon. There is a long list of other items, including more FC-4s.

Fibre Channel is not necessarily "the" interface of the 1990s, but along with ATM to cover the Wide Area Network (WAN) needs, Fibre Channel will be significant. As time goes on and prices come down, Fibre Channel will approach commodity status; it will start in the large systems and work-stations and work its way down toward personal computers.

BIBLIOGRAPHY

The Fibre Channel standard is made up of various parts, shown in Figure 20 (status as of the time of preparation of this book). New parts are being considered for addition to the set, to be developed later as industry support increases.

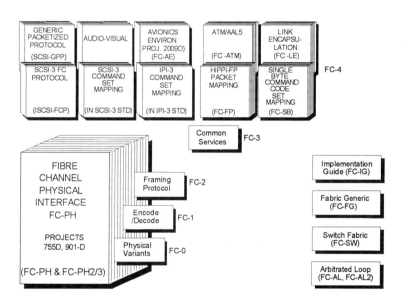

Figure 25. Fibre Channel Standard

The FC-PH standard is the base standard from which all others in the set are derived. There are standards for the topologies, the various command sets, and several protocols. Each part is a separate standard so that some parts become available in a reasonable period of time without waiting for unrelated parts to be finished. No one individual or group has an interest in all aspects of this wide ranging standard. By having the standard divided into individual parts, you use only those parts necessary and of interest to the product under consideration.

The Fibre Channel standard is divided into five levels. One or more standards are being developed to define each component of each level. The FC-PH standard defines levels FC-0, FC-1, and FC-2 and gives considerable guidance about the other levels. The FC-3 level currently has no functions defined for it. The FC-4 level describes mapping of individual channel or network protocols.

FIBRE CHANNEL STANDARD is published in individual documents (status as of August 1996):

FC-PH   Fibre Channel Physical Interface
            Standards: #X3.230-1994 (Rev.4.3)
FC-PH-2   Fibre Channel Enhanced Physical Interface
            Latest Draft: X3T11/Project 901/Rev. 7.4
            This document originally was called FC-EP
FC-PH-3   Fibre Channel Physical and Signaling Interface
            Latest Draft: XT311/Project 1119D/Rev.8.8
FC-IG   Fibre Channel Implementation Guide
            Latest Draft: X3T11/Project 956/Rev. 4.2
FC-FP   Fibre Channel Mapping to HIPPI-FP Framing Protocol
            Latest Draft: X3T11/Project 954/Rev.2.2
FC-LE   Fibre Channel, Link Encapsulation
            Latest Draft: X3T11/Project 955/Rev.1.0
            and X3T11/93-157 Internet - IPI&ARP on FC Ver.3.12
FC-SB   Single Byte Command Code Sets
            Latest Draft: X3.271-1996
GPP   SCSI-3 Generic Packetized Protocol
            X3T10/Project 991/Rev.8.0
FCP   SCSI-3 Fibre Channel Protocol
            Latest Draft: X3T10/Project 993D/Rev.8.0b
FC-FG   Fibre Channel Fabric Generic Requirements
            Latest Draft: X3T11/Project 958/Rev.3.4

FC-SW   Fibre Channel Switched Fabric Requirements
           Latest Draft: X3T11/Project 959/Rev.2.0
FC-AL   Fibre Channel Arbitrated Loop
           Latest Draft: X3T11/Project 960/Rev.4.5
FC-AL-2   X3T11/Project 1133 /Rev.5.1
FC-FLA   FL Loop Attachment Rev.2.1
FC-PLDA FC Private Loop Attach X3T11/Project 1162 / Rev.1.4
FC-ATM   Fibre Channel mapping to Asynchronous Transfer Mode,
           ANSI Project approved, no document yet.
FC-GS   Fibre Channel Generic Services
           Latest Draft: X3T11/Project 1050/Rev.3.0
FC-FLA   FC Loop Attachment/Rev.2.1
FC-PLDA FC Private Loop Attachment/X3T11/Project 1162/Rev.1.4
IP and ARP on FC - Pending RFC/X3T9.3/93-157
10-Bit Interface Specification - Technical Report X3T11/Project/1118/
                            Rev.2.3

The standards documents above are available from:

GLOBAL ENGINEERING DOCUMENTS
3130 South Harbor Blvd, Suite 330, Santa Ana, California 92704
(800)854-7179

Publications:

Fibre Channel, Volume 1: The Basics, written by Gary Stephens and
Jan Dedek, and published by ANCOT in 1995. This book includes over
250 pages of text and illustrations of the latest Fibre Channel concepts.
Order directly from ANCOT Corporation.

Fibre Channel Arbitrated Loop, written by Robert W. Kembel, the
Fibre Channel consultant, 1996. This is a detailed description of
Fibre Channel Arbitrated Loop operations.
Available directly from Connectivity Solutions in Arizona. See address
below.

Fibre Channel - Gigabit Communications and I/O for Computer Networks, written by Alan F. Benner, 1996, and published by McGraw-Hill, available in most bookstores.

Other information is available from:

FIBRE CHANNEL ASSOCIATION
12407 MoPac Expwy, North #100-357, Austin, Texas 78758
Tel: (512) 328-8422 or (800) 272-4618
Fax: (512) 328-8423 or (800) 524-8317

UNIVERSITY OF NEW HAMPSHIRE INTEROPERABILITY LAB
332 Morse Hall, Durham, New Hampshire 03824
(603) 862-1834

ENDL PUBLICATIONS
14426 Black Walnut Court, Saratoga, California 95070
(408)867-6630

The Fibre Channel Standard is being developed by:
ANSI Accredited Standards Committee
X3 Information Processing Systems
X3 Secretariat, IT (Information Technology)
1250 Eye Street NW, Suite 200, Washington DC 20005-3922
Tel: (202)737-8888 (press 1 twice) Fax: (202)638-4922 or 628-2829

Consulting services and education:

FSI Consulting Services (Mr. Gary Stephens)
1825 N. Norton Ave, Tucson, Arizona 85719
(520) 321-1725

Fibre Channel Group (Mr. Ed Frymoyer)
Box 398, Half Moon Bay, California 94019-0398
(415) 712-1986

Solution Technology (Mr. David Deming)
Box 104, Boulder Creek, California 95006
(408) 338-4285

Connectivity Solutions (Mr. Robert W. Kembel)
3061 N. Willow Creek Drive
Tucson, Arizona 85712
(520) 881-0877

GLOSSARY

8B/10B   a type of encoding and decoding bytes to reduce errors in transmission. This process is patented by IBM Corporation, which has agreed to a royalty-free, one-time documentation fee for use of its patent.

ANSI   American National Standards Institute

arbitrate   a process of selecting one L_Port from a collection of several ports that concurrently request use of the arbitrated loop.

arbitrated loop   a loop type of topology where two or more ports can be interconnected, but only two ports at a time can communicate

ATM   Asynchronous Transfer Mode network protocol

BER   Bit Error Rate. A measure of transmission accuracy. It is a ratio of bits received in error to bits sent.

bit   a single binary digit having a value of either 0 or 1. A bit is the smallest unit of data a computer can process.

buffer   a storage space temporarily reserved for a specific purpose. In Fibre Channel, a buffer is usually bigger than a single frame and may be the size of an entire sequence.

buffer-to-buffer   a method of transfer of information where the content is not known to the elements transferring the information

byte   an ordered set of 8 bits

CCITT   Consultative Committee International Telephone and Telegraph. An
                international association that sets worldwide communication standards,
                recently renamed International Telecommunications Union (ITU).

channel   an interface directed toward high speed transfer of large amounts of
                information

character   see transmission character.

coaxial cable   a cable made from a single copper wire clad with insulation and a shield,
                through which information is transmitted as electrical pulses. Used for
                high-speed transmission over short distances.

cyclic redundancy check (CRC)   an error detection mechanism that calculates a
                numeric value, using a special algorithm applied to a series of bytes and
                normally appended to the data. If the receiver executes the algorithm on the
                received data and no error has occurred, the newly generated CRC value
                should match the transmitted CRC value.

data   according to AT&T Bell Labs: "a representation of facts, concepts or
                instructions in a formalized manner, suitable for communication,
                interpretation, or processing." For Fibre Channel, data means the payload
                portion of a frame, which is not interpreted. Contrast with frame header.

decoding   a process to extract useful information from a string of bits

deserialization   the process of receiving information one bit at a time and reforming it
                into a larger unit, such as a transmission character or a byte

destination address   a value in the frame header of each frame that identifies the port in
                a node that is to receive the frame

device   see node and peripheral device.

disassembly a process of dividing a source buffer into payloads for transmission in
                frames

64

dual-simplex a communication protocol that permits simultaneous transmission in both directions in a link, usually with no flow control

encoding a process to change the form of information into an alternate form, such as changing handwritten letters on a page to bytes in a computer

end-of-frame delimiter (EOF) a delimiter used to mark the end of a frame

exchange a set of one or more sequences, usually associated with an I/O operation, having the same exchange identifier

exchange identifier the numeric value assigned to an exchange

F_Port a port in a fabric where an N_Port or NL_Port may attach

fabric a group of interconnections between ports that includes a fabric element

FCP Fibre Channel Protocol

fiber a wire or optical strand. Spelled "fibre" in the name Fibre Channel.

fiber optic cable jacketed cable made from thin strands of glass through which light pulses transmit data. Used for high-speed transmission over medium to long distances.

field the smallest unit of data to hold application information. A field may be a single bit or many bytes. For example, "name" is a field in a mailing address.

FL_Port a port in a fabric where an N_Port or an NL_Port may attach

frame an indivisible unit for transfer of information in Fibre Channel

frame header the first field in the frame content containing addresses and other control information about the frame

full duplex a communication protocol that permits simultaneous transmission in both directions, usually with flow control

giga   a prefix that means a billion or a thousand million units, as in gigabaud and gigabyte

GPP   Generic Packetized Protocol

half duplex   a communication protocol that permits transmission in both directions, but in only one direction at a time

HIPPI   High Performance Parallel Interface. An ANSI standard for high-speed transfer of information in a dual-simplex manner over a short parallel bus.

host   a processor, usually a CPU and memory. Typically, a host communicates with peripheral devices over an interface to carry out I/O operations.

I/O   input/output

Idle   a fill word transmitted on a fiber to fill the space between frames. In Fibre Channel, a fiber must have continuous transmission words sent and received at regular intervals for it to remain operational.

inbound fiber   one fiber in a link that carries information into a port

interface   a set of protocols used between components, such as cables, connectors, and signal levels

IP   Internet Protocol. A set of protocols developed by the Department of Defense to communicate between dissimilar computers across networks.

IPI   Intelligent Peripheral Interface. An ANSI standard for controlling peripheral devices by a host computer.

L_Port   an arbitrated loop port: either an NL_Port, an FL_Port, or a GL_Port

LAN   Local Area Network

laser   Light Amplification by Stimulated Emission of Radiation. A device for generating coherent radiation in the visible, ultraviolet, and infrared portions of the electromagnetic spectrum.

LED   Light-emitting diode

link   one inbound fiber and one outbound fiber connected to a port

micron   millionth of a meter - a micrometer (μm)

multimode fiber   an optical waveguide which allows more than one mode (rays of light) to be guided

N_Port   a port attached to a node for use with point-to-point or fabric topology

NL_Port   a port attached to a node for use in all three topologies

network   an arrangement of nodes and connecting branches. Also, a configuration of data processing devices and software connected for information exchange.

node   a device that has at least one N_Port or NL_Port

offline   something not active or not available for access in a system

online   something active or available for access in a system

Open Fibre Control (OFC)   a safety interlock system that controls the optical power level of an open optical fiber cable

optical fiber   any filament of fiber, made of dielectic material, that guides light

originator   the logical function associated with an N_Port responsible for originating an exchange

outbound fiber   one fiber in a link that carries information away from a port

parallel transmission   the simultaneous transmission of bits over multiple fibers. Accomplished by devoting a fiber for each bit. Parallel data transmission is very fast, but usually is practical only for short distances (typically under 500 feet) because of signal recovery and signal quality problems. Contrast with serial transmission.

payload   that portion of the data field in a frame left after removing any optional headers

point-to-point   a topology where exactly two ports communicate

port   an access point in a device where a link attaches. See N_Port, NL_Port, F_Port, and FL_Port.

primitive sequence   a transmission word defined by Fibre Channel that conveys special control information between ports on a link. A primitive sequence is recognized when three adjacent transmission words of the same value are received.

primitive signal   a transmission word defined by Fibre Channel that conveys control information between ports on a link. A primitive signal is recognized when only one transmission word is received.

protocol   a convention for data transmission that defines timing, control, format and data representation

reassembly   a process of reordering and extracting the payload fields from frames in a sequence to reconstruct a copy of a source buffer in a destination buffer

receiver   the circuitry that receives signals on a fiber. Also, the ultimate destination of data transmission.

responder   the logical function in an N_Port responsible for supporting the exchange initiated by the originator in another N_Port

SCSI   Small Computer System Interface. An ANSI standard for controlling peripheral devices by one or more host computers. Versions are SCSI-1 (obsolete), SCSI-2, and SCSI-3.

sequence   a set of one or more frames identified as a unit within an exchange.

sequence count   a value in a frame header that helps identify the order in which frames were transmitted

sequence identifier   the field in a frame header that sets one sequence apart from another between a pair of terminal N*_Ports

serial transmission   data communication mode where bits are sent in sequence in a single fiber. Contrast with parallel transmission.

serialization   a process of sending transmission characters one bit at a time by serial transmission

single-mode fiber   a step index fiber waveguide in which only one mode (ray of light) will propagate above the cut-off wavelength

source address   a field in a frame header that identifies its sender

start-of-frame delimiter (SOF)   a delimiter used to mark the beginning of a frame

switch   the name of an implementation of the fabric topology

topology   the components used to connect two or more ports together. Also a specific way of connecting those components, as in point-to-point topology.

transceiver   a transmitter/receiver module

transfer rate   the rate at which bytes or bits are transferred, as in megabytes per second.

transmission character   the 10-bit encoded form of a byte or special code using the 8B/10B encoding algorithm

transmission word   a set of four transmission characters, 40 bits. A transmission word is the smallest information unit transmitted in Fibre Channel.

twinax   a cable made from two parallel copper wires clad with insulation and a shield, through which information is transmitted as electrical pulses. Used for high-speed transmission over short distances.

WAN   Wide Area Network

wavelength   a linear value corresponding to a specific frequency

INDEX

ANCOT
CORPORATION

Model
FCA-5000
Fibre Channel Analyzer

HIGHLIGHTS

▼ **two independent analyzer channels with own trigger capability (two analyzers in one box)**

▼ **split screen display to monitor both channels at once**

▼ **transfer rates: 1Gbps, 531Mbps, or 266Mbps**

▼ **media: optical fiber or copper (twisted pair)**

▼ **topology: fabric, arbitrated loop, or point-to-point**

▼ **connects to tested link using standard SC (optical fiber) or DB-9(copper) connectors**

The ANCOT Fibre Channel Analyzer is a tool for development, integration, and repair of Fibre Channel based systems. This instrument attaches non-intrusively to both fibres (In & Out) of the fibre link to monitor and record all activity. When recording is stopped, it can play back and portray on the screen every facet of information about activity on the fibre link. The display format is in plain English for ease of understanding. Packaged in a transportable enclosure, it is well suited for work in the laboratory as well as for field applications where versatility and high reliability are important.

ADDITIONAL SPECIFICATIONS:

▼ circular trace memory for each channel: 128k events deep is standard (512k will be optional in future revisions)
▼ triggering capability:
  stop recording if trace memory is full,
  stop/start trigger circuit uses four (or eight optionally) sliding one-word comparators (0/1/X), one set on each fibre of a FC link. The setup employs simple binary equations (or high level logical selections - in future revisions). Four-level trigger is standard, eight-level is optional. External trigger can be brought in via a BNC connector on the back panel
▼ passive, non-intrusive tracing on both fibres of a link simultaneously
▼ both directions of traffic displayed in time-synchronized split-screen format
▼ fibre traffic is captured to the trace memory directly by hardware
▼ traced data is saved in "raw" (10B) format to allow capture of illegal codes and to enable protocol upgrades in the future
▼ display formats: Structured format (interpreted in English), and Low level formats (either Raw, 'K/D'type interpreted, or combination with Hex and ASCII)
▼ data compression to allow efficient utilization of trace memory when recording repeating patterns (e.g. 'Idles')
▼ Start - Pause - Stop tracing capability
▼ built-in Electroluminescent (EL) display and external keyboard
▼ embedded operating system
▼ user commands by single keystroke or menu selections
▼ hard copy capability using external parallel printer
▼ remote control using RS-232 serial port PC-AT compatible, with selectable baud rate (up to 115k), data format and parity
▼ trace offload and download to/from a host via serial port

PACKAGING & POWER

Housed in an elegant, high quality, metal, fan cooled enclosure
Dimensions : 16"(W) x 7.5"(H) x 14"(D) Weight : 23 lb
Power: Built-in autoranging switching power supply 110/220V AC, 50-60Hz

PLANNED ENHACEMENTS

▼ Trace memory will be available for higher capacities
▼ Display in structured format will have FC-4 level payload interpretation for SCSI, IP and other protocols
▼ Filtering and network management ('Sniffer') options will be available

ORDERING INFORMATION

The base unit comes with 128k events trace memory, built in flat panel screen, and external keyboard. The base unit does not include the pod, trigger or other options. When ordering, specify media type, transfer rate,four or eight-level trigger options, and cables.

For availability or other information call the factory

TRACE DISPLAY

The Fibre Channel link uses two fibres - one In and one Out. The display shows both channels side by side on a split screen. Two display modes are available: the Structured and the Low level. Powerfull search function in both directions is available for analyzing longer traces.

STRUCTURED DISPLAY mode shows activity for each channel on a split screen in 'readable English'. The data part is shown as payload with byte count. Payload can be interpreted depending on the type of protocol mapped (e.g. SCSI3/FCP,..) 'Idles' are compressed and are shown as time period.

```
000000:Idle 21.9 us                      Idle 21.9 us
000002:    SOFi2                          Idle 2.5 us
000003:          R_CTL:06    D_ID:000002  .
000004:                      S_ID:000001  .
000005:          Type:08(FCP)    F_CTL:390008
000006:          Seq:00, DF_CTL:00, #0000  .
000007:          OX_ID:C000 RX_ID:FFFF    .
000008:          Parameter:00000000       .
000009:               Payload:32($20) Bytes  .
00000A:          ·CRC:55E0B14B            .
000011:    EOFn                           .
000012:Idle 8.0 us                        Idle 8.0 us
000014:Idle 160 ns                        R_RDY
000015:Idle 14.0 ms                       Idle 14.0 ms
000017:Idle 1.4 us                        SOFn2
000018:    .                                    R_CTL:C0    D_ID:000001
000019:    .                                                S_ID:000002
00001A:    .                                    Type:00        F_CTL:F90008
00001B:    .                                    Seq:00, DF_CTL:00, #0000
00001C:    .                                    OX_ID:C000 RX_ID:C040
00001D:    .                                    Parameter:00000000
00001E:    .                                         Payload:0($0) Bytes
00001F:    .                                    CRC:FCD96011
000020:Idle 8.0 us
```
————— Channel A —————   ————— Channel B —————

LOW LEVEL DISPLAY mode shows activity in four variations: the 'K/D' type interpreted, in combinations with Hex for frame header and Hex or ASCII for payload data. Each line contains actual data with primitive meaning of the word displayed (e.g. R_RDY, SOFn1, etc.), and disparity (+, -, or ~). In case of the 'Idle' words, only one line is displayed for however many 'Idle's appear in a row, along with the count.

```
000000:K28.5-D21.4+D21.5~D21.5~ Idle   ≥ K28.5-D21.4+D21.5~ Idle
000001:   Repeated 00 0000 0089         ≥   Repeated 00 0000 0089
000002:K28.5-D21.5-D21.2-D21.2- SOFi2   ≥ K28.5-D21.4+D21.5~D21.5~ Idle
000003: 06    00    00    02       F Hdr  ≥ K28.5-D21.4+D21.5~D21.5~ Idle
000004: 00    00    00    01               ≥ K28.5-D21.4+D21.5~D21.5~ Idle
000005: 08    39    00    08               ≥ K28.5-D21.4+D21.5~D21.5~ Idle
000006: 00    00    00    00               ≥ K28.5-D21.4+D21.5~D21.5~ Idle
000007: C0    00    FF    FF               ≥ K28.5-D21.4+D21.5~D21.5~ Idle
000008: 00    00    00    00               ≥ K28.5-D21.4+D21.5~D21.5~ Idle
000009:   Repeated 00 0000 0002         ≥   Repeated 00 0000 0002
00000A: 00    00    00    01       Payld  ≥ K28.5-D21.4+D21.5~D21.5~ Idle
00000B: 0A    00    37    D9               ≥ K28.5-D21.4+D21.5~D21.5~ Idle
00000C: 04    00    00    00               ≥ K28.5-D21.4+D21.5~D21.5~ Idle
00000D: 00    00    00    00               ≥ K28.5-D21.4+D21.5~D21.5~ Idle
00000E: 00    00    00    00               ≥ K28.5-D21.4+D21.5~D21.5~ Idle
00000F: 00    08    00    00               ≥ K28.5-D21.4+D21.5~D21.5~ Idle
000010: 55    E0    B1    4B       CRC    ≥ K28.5-D21.4+D21.5~D21.5~ Idle
000011:K28.5+D21.4-D21.6-D21.6~ EOFn     ≥ K28.5-D21.4+D21.5~D21.5~ Idle
000012:K28.5+D21.4-D21.5~D21.5~ Idle     ≥ K28.5-D21.4+D21.5~D21.5~ Idle
000013:   Repeated 00 0000 0032         ≥   Repeated 00 0000 0032
000014:K28.5+D21.4-D21.5~D21.5~ Idle     ≥ K28.5-D21.4+D10.2-D10.2~ R_RDY
000015:K28.5+D21.4-D21.5~D21.5~ Idle     ≥ K28.5-D21.4+D21.5~D21.5~ Idle
000016:   Repeated 00 0001 55CD         ≥   Repeated 00 0001 55CD
000017:K28.5+D21.4-D21.5~D21.5~ Idle     ≥ K28.5-D21.5-D21.1-D21.1~ SOFn2
```
————— Channel A —————   ————— Channel B —————

Specifications subject to change without notice   02/96

☎ (415) 322-5322
Fax: (415) 322-0455
http://www.ancot.com

ANCOT
CORPORATION

115 Constitution Drive
Menlo Park, California
94025

Model
FSB-8001/8010
Fibre Channel to SCSI Bridge
266Mbps / 1Gbps

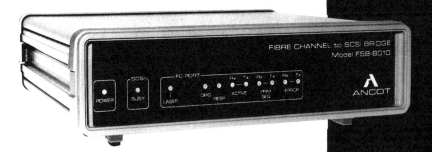

HIGHLIGHTS

▼ allows interconnection of SCSI devices and Fibre Channel networks

▼ automatic operation, no programming needed

▼ 5.25 inch form factor, 9 inch deep

▼ FIBRE CHANNEL port:
  - ▼ standard transfer rates: 266Mbps or 1Gbps
  - ▼ media: optical fiber, or copper
  - ▼ topologies: Point-to-Point, Fabric, or Arbitrated Loop
  - ▼ classes of service: class 1, 2, and 3

▼ PARALLEL SCSI-2/3 port:
  - ▼ WIDE & FAST/10 or FAST/20 ('UltraSCSI'), using 68-pin HD 'P' connector
  - ▼ SCAM protocol tolerant
  - ▼ Single-Ended or optionally Differential interface
  - ▼ built-in SCSI bus termination and TERMPWR

The FSB-80xx is a device used for connecting parallel SCSI devices to Fibre Channel networks. The operation of the bridge is bidirectional - it can act as Fibre Channel originator or responder, or as SCSI initiator or target. Internally, it uses "store & forward" scheme for moving data between the Fibre Channel and SCSI. The SCSI protocol is mapped to the Fibre Channel using SCSI3/FCP. In future releases, faster protocol will also be offered. The bridges can be used for extending SCSI bus over very long distances. With two FSB-8001 bridges connected back to back over an optical fiber link, distance between the SCSI devices can be extended to 2km or more.

Fibre Channel to SCSI Bridge   Model FSB-8001/8010

FIBRE CHANNEL PORT

▼ protocols: FC-PH Rev:4.3, SCSI-3/FCP Rev:12.0, FC-AL Rev:4.5(1Gbps only)
▼ transfer rates: 266Mbps(FSB-8001) or 1Gbps(FSB-8010)
▼ fiber interface: 1300nm LED Multimode standard, other types optional
▼ media: optical fiber or copper
▼ topologies: Point-to-Point, Fabric, or Arbitrated Loop (1Gbps only)
▼ classes of service: class 1, 2, and 3
▼ connector: SC for optical fiber, or DB-9 for copper

SCSI PORT

▼ responds as either: SCSI Initiator or Target
▼ SCSI-2/3 FAST/10 or FAST/20 ('UltraSCSI'), WIDE (2-bytes) using 68-pin HD 'P' type connector
▼ SCSI controller chip is the 53C770
▼ Single-Ended(S-E) interface is standard. Optional Differential (DF) interface is installed as a plug-in daughter board.
▼ On board termination, enabled by DIP switch or in S/W. The S-E version uses built-in active terminator, the DF version uses standard resistor network type.
▼ programmable SCSI ID
▼ command sets supported: SCSI-3 SPC, SBC, SSC, SMC

OTHER

▼ embedded operating system
▼ internal MPU is the 68340/25MHz
▼ firmware stored in flash EPROM allows easy updates/upgrades over serial line
▼ separate transmit & receive data buffers, 128KBytes each. Buffers are dual ported, with one port dedicated exclusively to the FC port, the other port shared between the internal MPU and SCSI chip.
▼ data transfer uses 'store & forward' scheme. Data is written to the buffer only once; internal transfer is managed using pointers and DMA hardware.
▼ all speed critical functions are implemented in autonomous sequencers operating independently, communicating with the internal MPU by interrupts.
▼ simple setup (via serial port) may be required during initial installation

USING TWO BRIDGES AS A SCSI EXTENDER

Two bridges can be connected back to back over an optical fiber link to be used as a SCSI bus extender. Such an extender will be transparent to the hosts or devices connected. Initialization is internal, executed only between the two connected Fibre Channel ports. On power up, the SCSI ports on both sides scan their devices. The sides inform each other, then they remember both configurations. The SCSI devices on one side do not know how they are connected to devices on the other SCSI side, they act as if both sides were connected by a single, although somewhat longer, SCSI cable.

SAFETY AND RADIATION TESTING & CERTIFICATION

All units (except for the introductory lot and prototypes) will be tested to all standard specifications - the UL, TUV, FCC, including the CE required in Europe.

MTBF

The bridge is designed with MTBF requirements in mind, with the intention to achieve MTBF as high as possible.   All circuitry is assembled on a single PCB for high reliability. There are no moving parts or other components requiring maintenance or calibration.

PACKAGING & POWER

Housed in an elegant, aluminum, stand-alone, 5.25" form factor enclosure.

DIMENSIONS :

FSB-8001
5.75"(W) x 1.625"(H) x 9"(D)
WEIGHT : < 4 lb
FSB-8010
10"(W) x 2.75"(H) x 12"(D)
WEIGHT : < 4.5 lb
POWER(both models) : regulated +5V DC power supply for 110-220V 50-60Hz. The external wallplug type (included).

WARRANTY

One year, return to factory.

ORDERING INFORMATION

FSB-8001/266SF version w.266Mbps,1300nm LED, optical fiber(MM, 50/125), S-E SCSI
FSB-8001/266DF same as the above except with Differential SCSI
FSB-8010/1GSF version w.1Gps, optical fiber (SC conn.) (MM, 50/125), S-E SCSI
FSB-8010/1GDF same as the above except with Differential SCSI

 (415) 322-5322
Fax: (415) 322-0455
http://www.ancot.com

ANCOT
CORPORATION

115 Constitution Drive
Menlo Park, California
94025